AN ILLUSTRATED SURVEY OF RAILWAY STATIONS BETWEEN SOUTHPORT & LIVERPOOL

1848-1986

COMPILED BY ROB GELL

HEYDAY PUBLISHING COMPANY

DEDICATED TO DOUGLAS THOMPSON

Few people realise that 19th century Southport was the third biggest seaside resort in the country being surpassed only by Brighton and Great Yarmouth. Railways considerably aided the growth of Southport as it succeeded the bathing resort of Liverpool and the spa town of Ormskirk. The first line to the area was opened by the Liverpool, Crosby and Southport Railway in 1848 terminating at Eastbank Street (the extension to Chapel Street opening three years later). A route from Manchester (via Burscough) followed in 1855 and the effect of these lines upon the number of visitors to the town was staggering. The line from Preston first reached the town in 1878 and was followed by the Cheshire Lines Committees course (via Altcar) in 1884. The Barton branch of 1887 completed the areas rail network bringing the number of iron roads to the town to five. The line through Ormskirk to Burscough and beyond was opened in 1849 providing another important trade and commuter link. As would be expected the areas between Southport, Liverpool and Ormskirk also benefitted from the coming of the railway. The farming communities around Halsall and Ormskirk were now able to convey produce to markets further afield at more competitive prices. Faster commuting saw areas like Formby and Ainsdale become more developed and the whole of the region hosted more day-trippers as the need for overnight accommodation dwindled. The 20th century saw Southport gradually become more conservative towards tourism and Blackpool was allowed to succeed it as the regions major seaside resort. With road improvement motor transport ate into railway receipts and the all too familiar outcome was the closure of three of the towns rail links. The Barton branch was the first to go when passenger facilities were severed in 1938. The route from Aintree (via Woodvale) went the same way in 1952 and the Preston line was axed in 1964. Ormskirk is now a double ended terminus having lost its through service and the route from here to Preston has been reduced to single track. The following pages table each of the areas stations since 1848 together with photographs, useful dates and other information. A map is also included – see centre pages.

Grateful thanks are extended to the following:

P. Bolger	F. Dean	Lens of Sutton	J. Roles
Mrs. J. Burgess	P. Gibb	L.G.R.P.	J. Ryan
H. C. Casserley	J. Gilmour	C. Loker	Southport Library
Mr. & Mrs. Cawfield	B. Hilton	B. Mathews	D. Thompson
Crosby Library	J. B. Hodgson	J. Peden	J. Ward
G. Danby	Knowsley Library	H. B. Priestley	P. Woolley

ISBN 0 947562 04 4

First Published 1986

HEYDAY PUBLISHING COMPANY
KINGSWAY HOUSE, 83 CROSBY ROAD NORTH,
CROSBY, MERSEYSIDE, L22 4QD, ENGLAND

Ainsdale (Lancashire & Yorkshire Railway) *Line opened* 21-7-1848 *Station opened* 21-7-1848 Still open. Single line until 1852 being electrified in 1904.

Ainsdale (LYR) station circa 1910 (Photo C. Loker collection)

Ainsdale Beach (Cheshire Lines Committee) *Line opened* 1-9-1884 *Station opened* 19-6-1901 as 'Seaside' It was renamed 'Ainsdale Beach' on 1-1-1912 *Station closed* 1-1-1917 *Station re-opened* 1-4-1919 *Station closed* 7-1-1952. *Line closed* 7-7-1952 The line was lifted shortly after and although nothing now remains of the station the trackbed has since been utilised as the coastal road between Woodvale & Southport. The station was situated near the holiday camp North of the present-day roundabout.

A 1949 view of Ainsdale Beach (CLC) looking North (Photo D. Thompson)

Altcar & Hillhouse (Cheshire Lines Committee) *Line opened* 1-9-1884 *Station opened* 1-9-1884 *Station closed* 1-1-1917 *Station re-opened* 1-4-1919 *Station closed* 7-1-1952 *Line closed* 7-7-1952. The line North of the station was lifted shortly after but rail access from Liverpool was retained up to 1960 to serve private sidings at the site. This station once hosted trains from the LYR Barton branch via the nearby Hillhouse Junction but these were discontinued in 1926 (The 'Altcar Bob' service was so named because it terminated here up to this time – see Barton notes). The station was situated on the South side of the present Formby – Downholland road (B5195) near the sewage works. Platform mounds are still visible and the road overbridge remains.

Altcar & Hillhouse (CLC) in 1949 looking South (Photo D. Thompson)

Altcar Rifle Range (Lancashire & Yorkshire Railway) *Line opened* 21-7-1848 *Station opened* 1862 as 'Hightown Rifle Station' It was renamed in 7-1886 *Station closed* circa 10-1893 *Station re-opened* circa 4-1894 *Station closed* 3-10-1921 The line was electrified in 1904. The station was situated a third of a mile North of Hightown. A small tramway plied between the firing ranges and the bay platform with munitions – see plan.

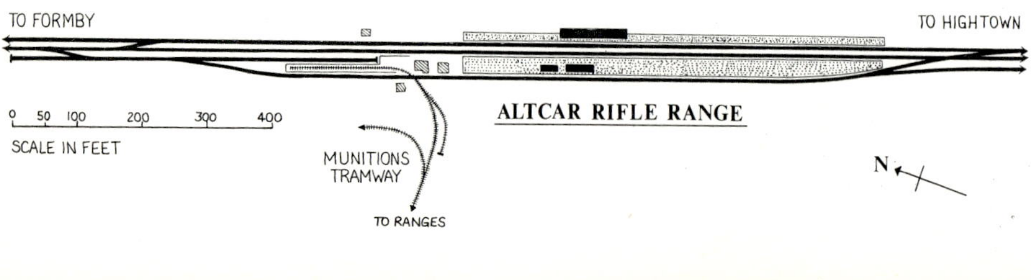

Aughton Park (Lancashire & Yorkshire Railway) *Line opened* 2-4-1849 *Station opened* 5-1907 Still open. Electrification reached Aughton in 1913 when it is most likely that the station was equipped with proper platforms. Prior to this an economical rail-motor operated between Aintree & Ormskirk – see Barton for photo of identical conveyance.

Aughton Park prior to 1913 (Photo Heyday Publishing Company)

Aughton Park as viewed from the roadbridge in 1955 with Crooks sand quarry siding visible in the distance (Photo H. B. Priestley)

Barton (Lancashire & Yorkshire Railway) *Line opened* 1-9-1887 *Station opened* 1-11-1887 as 'Barton' It was renamed 'Downholland' on 2-6-1924 *Station closed* 26-9-1938 *Line closed* 21-1-1952 and tracks lifted shortly afterwards. Although this was mainly the terminus for trains on the old Liverpool, Southport & Preston Junction Railway line from Southport, through trains to Altcar & Hillhouse (CLC) were operated until 1926. The station stood on Station Road which is off the A567 by the 'Blue Bell' public house. The road overbridge still stands and platform remains are visible today. The Barton branch was notable for the 'Altcar Bob' service, introduced in July 1906 – a diminutive steam loco and solo coach (see photo). The vehicle did not require turning because of remote controls for the driver located at the rear of the coach. The origin of the term 'Bob' is uncertain. Old railwaymen state it was named after one of the original drivers but others insist it was a term given to many small locomotives. The latter is more likely, however, ex-patrons suggest it related to the cost of a journey in early days – 'Bob' = shilling (5 new pence).

Looking South to Barton in 1949 (Photo D. Thompson)

An ancient view of Barton looking North (Photo J. Gilmour collection)

One of the 'Altcar Bob' engines at Barton circa 1930 — the loco is LMS No 10600 (ex LYR No. 3) (Photo J. Gilmour collection)

A group of young employees around the turn of the century (Photo J. Gilmour collection)

Bescar Lane (Lancashire & Yorkshire Railway) *Line opened* 9-4-1855 *Station opened* 9-4-1855 Still open. This site had the distinction of being the lowest station on the LYR network – 12½ feet above sea-level.

Loco No 45061 stopped at Bescar Lane with a Manchester bound train in 1949 (Photo D. Thompson)

Birkdale (Lancashire & Yorkshire Railway) *Line opened* 21-7-1848 *Station opened* 1851 Still open. Renamed 'Birkdale Park' 4-1854 but reverted to 'Birkdale' about 4-1965. The line was single track until 1852 and electrification came in 1904.

Looking North along Birkdales platforms (LYR) in 1949 (Photo D. Thompson)

Birkdale Palace (Cheshire Lines Committee) *Line opened* 1-9-1884 *Station opened* 1-9-1884 *Station closed* 1-1-1917 *Station re-opened* 1-4-1919 *Station closed* 7-1-1952 *Line closed* 7-7-1952 The line was lifted shortly after and nothing now remains of the station. The trackbed has since been utilised as the coastal road between Woodvale and Southport (although this road diverges from the original rail route immediately South of the Palace station site). The station was situated between Oxford Road and Weld Road with access from the latter.

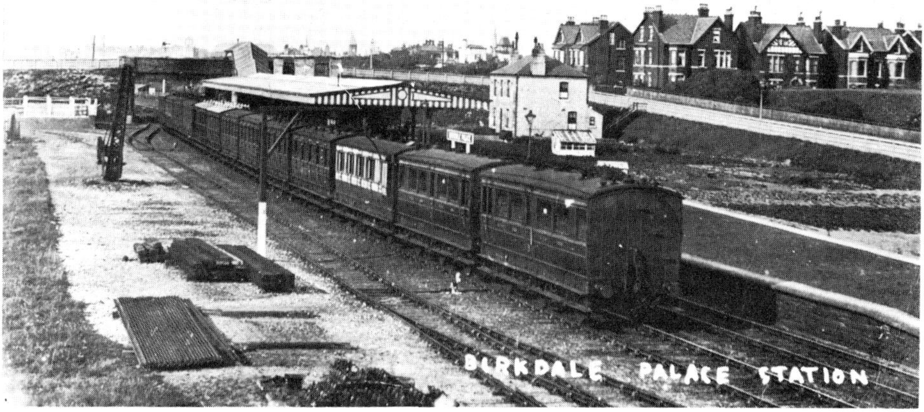

A circa 1910 view of Birkdale Palace (CLC) from Oxford Road (Photo Heyday Publishing Company)

An aerial view of Birkdale Palace (CLC) after closure (Photo J. Gilmour collection)

Blowick (Lancashire & Yorkshire Railway) *Line opened* 9-4-1855 *Station opened* early 1871 as 'Cop End' being renamed 'Blowick' on 1-10-1871 *Station closed* 25-9-1939 *Line closed* 14-6-1965 and tracks lifted shortly afterwards. The closure of this part of the Southport – Wigan line (St. Lukes Stn. to Pool Hey Junc.) forced trains operating this service to divert via Meols Cop, a practice that still exists. Blowick was situated at a level crossing on Meols Cop Road North of the school.

Blowick circa 1910 with the ex-Ormskirk railmotor approaching. It will be seen that the platforms were either side of the level crossing (Photo P. Bolger collection)

Blundellsands & Crosby (Lancashire & Yorkshire Railway) *Line opened* 21-7-1848 *Station opened* 21-7-1848 as 'Crosby' It was renamed 'Crosby & Blundellsands' on 1-6-1852 and then 'Blundellsands & Crosby' in June 1878. Still open. The track was single until 1852 and the station was originally sited at a level crossing on Mersey Road (see photo). In 1881 the station was moved to its present site and a road-bridge was erected on Mersey Road replacing the level crossing. The line was electrified in 1904.

The exterior of Blundellsands & Crosby in Edwardian days — note the enamel adverts and horse-drawn vehicles (Photo Heyday Publishing Company)

Blundellsands & Crosby (Liverpool platform) circa 1908 (Photo Heyday Publishing Company)

The original Blundellsands & Crosby station and level crossing at Mersey Road circa 1879. The loco is LYR No. 28 (2-4-0 well tank) and was rebuilt as a saddle-tank circa 1880 being withdrawn from service in December 1885 (Photo Crosby Library)

Burscough Bridge (Lancashire & Yorkshire Railway) *Line opened* 9-4-1855 *Station opened* 9-4-1855 Still open.

Burscough Bridge looking East in early days (Photo P. Gibb collection)

Burscough Bridge looking West (Photo D. Thompson)

Burscough Junction (Lancashire & Yorkshire Railway) *Line opened* 2-4-1849 *Station opened* 2-4-1849 Still open although the station is now single line and little more than a bus-stop owing to the reduction of services including the cessation of Southport – Ormskirk trains.

Looking North to Burscough Junction in 1955 with loco No. 45558 on a Preston – Liverpool train (Photo H. B. Priestley)

A closer view of the buildings at Burscough Junction (Photo D. Thompson)

Butts Lane Halt (Lancashire & Yorkshire Railway) *Line opened* 1-9-1887 *Station opened* 3-1907 *Station closed* 26-9-1938 Line still open although nothing remains of this platform-less halt. As the name implies it stood on Butts Lane and access was from the South side of the bridge. Once part of the Barton branch, the line now carries diverted Southport – Wigan services owing to the closure of the Pool Hey Junc. – St. Lukes section (see also Plex Moss Lane view and Barton notes).

Formby (Lancashire & Yorkshire Railway) *Line opened* 21-7-1848 *Station opened* 21-7-1848 as 'Formby & Altcar' It was renamed 'Formby' circa 1-1866 The track was single until 1852 and electrification came in 1904. Still open.

Formby station prior to the provision of the road overbridge which replaced the level crossing visible here (Photo P. Woolley collection)

Formby Power Station (Lancashire & Yorkshire Railway) *Line opened* 21-7-1848 *Station opened* circa 1917' *Station closed* circa 1944 Line still open. This small Halt was opened for workers at the nearby Power Station (a Railway installation which generated current for the Liverpool–Southport electrics). It stood South of Hogshill Lane and just North of the River Alt.

Freshfield (Lancashire & Yorkshire Railway) *Line opened* 21-7-1848 *Station opened* circa 4-1854 Still open. The line was electrified in 1904.

A Northerly view of Freshfield from the footbridge in Edwardian days (Photo B. Mathews collection)

Hall Road (Lancashire & Yorkshire Railway) *Line opened* 21-7-1848 *Station opened* 1874 Still open. Electrification was introduced in 1904. This station was opened as a result of the single-mindedness of one Joseph Gardner, a rich merchant of the area. He asked the railway company to build it thus saving him a long walk to Crosby from his isolated home. Predictably the LYR replied that another five houses would have to exist before consideration could be given to such a request. Mr. Gardner duly had them built and a stunned LYR conceded defeat! Hall Road is also notable for a tragedy that took place on 27-7-1905 – a Southport bound express hit an empty train and 21 people lost their lives with a further 45 injured – see photo.

Hall Road looking North from the footbridge (Photo Lens of Sutton)

Part of the wreckage from the 1905 disaster showing wheels from a destroyed car driven underneath the standing vehicle (Photo P. Woolley collection)

Halsall (Lancashire & Yorkshire Railway) *Line opened* 1-9-1887 *Station opened* 1-11-1887 *Station closed* 26-9-1938 *Line closed* 21-1-1952 and the tracks were lifted shortly afterwards. This ex-Liverpool, Southport & Preston Junction Railway station stood North of Carr Moss Lane and the building has survived as a dwelling conversion. The 'Altcar Bob' once operated through this station – see Barton notes.

Looking South to Halsall circa 1910 (Photo J. B. Hodgson collection)

Staff posing on the Southport bound platform (Photo Mrs. J. Burgess)

A close-up of the station buildings at Halsall (Photo Mrs. J. Burgess)

Heathey Lane Halt (Lancashire & Yorkshire Railway) *Line opened* 1-9-1887 *Station opened* 3-1907 *Station closed* 26-9-1938 *Line closed* 21-1-1952 The track remained until 1964 for the storage of excursion stock. There were no platforms and access was via steps on the North side of the Heathey Lane (B5243) roadbridge which remains today (see also Plex Moss Lane view and Barton notes).

Hightown (Lancashire & Yorkshire Railway) *Line opened* 21-7-1848 *Station opened* 21-7-1848 Still open. The track was single until 1852 and electrification was introduced in 1904.

An early view of Hightown looking South (Photo Heyday Publishing Company)

Hillside (London Midland & Scottish Railway) *Line opened* 21-7-1848 *Station opened* 5-1926 Still open.

Hillside looking North (Photo Author)

Kew Gardens (Lancashire & Yorkshire Railway) *Line opened* 1-9-1887 *Station opened* 1-11-1887 *Station closed* 26-9-1938 *Line closed* 21-1-1952 The tracks remained until 1964 for the storage of excursion stock. This ex-Liverpool, Southport & Preston Junction Railway station stood North of the A570 between Meols Cop Road and Foul Lane. The site has been reclaimed but one of the nearby bridge supports can still be seen. The station was named after the nearby 12 acre park and boating lake which it served (closed about 1930). See Barton notes also.

Kew Gardens looking towards Southport (Photo J. Gilmour collection)

Kirkby (Lancashire & Yorkshire Railway) *Line opened* 20-11-1848 *Station opened* 20-11-1848 Still open although considerably rebuilt.

Kirkby circa 1910 (Photo Heyday Publishing Company)

This view of Kirkby taken about the same time shows the opposite platform and goods bay (Photo Knowsley Library)

Railway Stations between

SOUTHPORT & LIVERPOOL

1848 - 1986

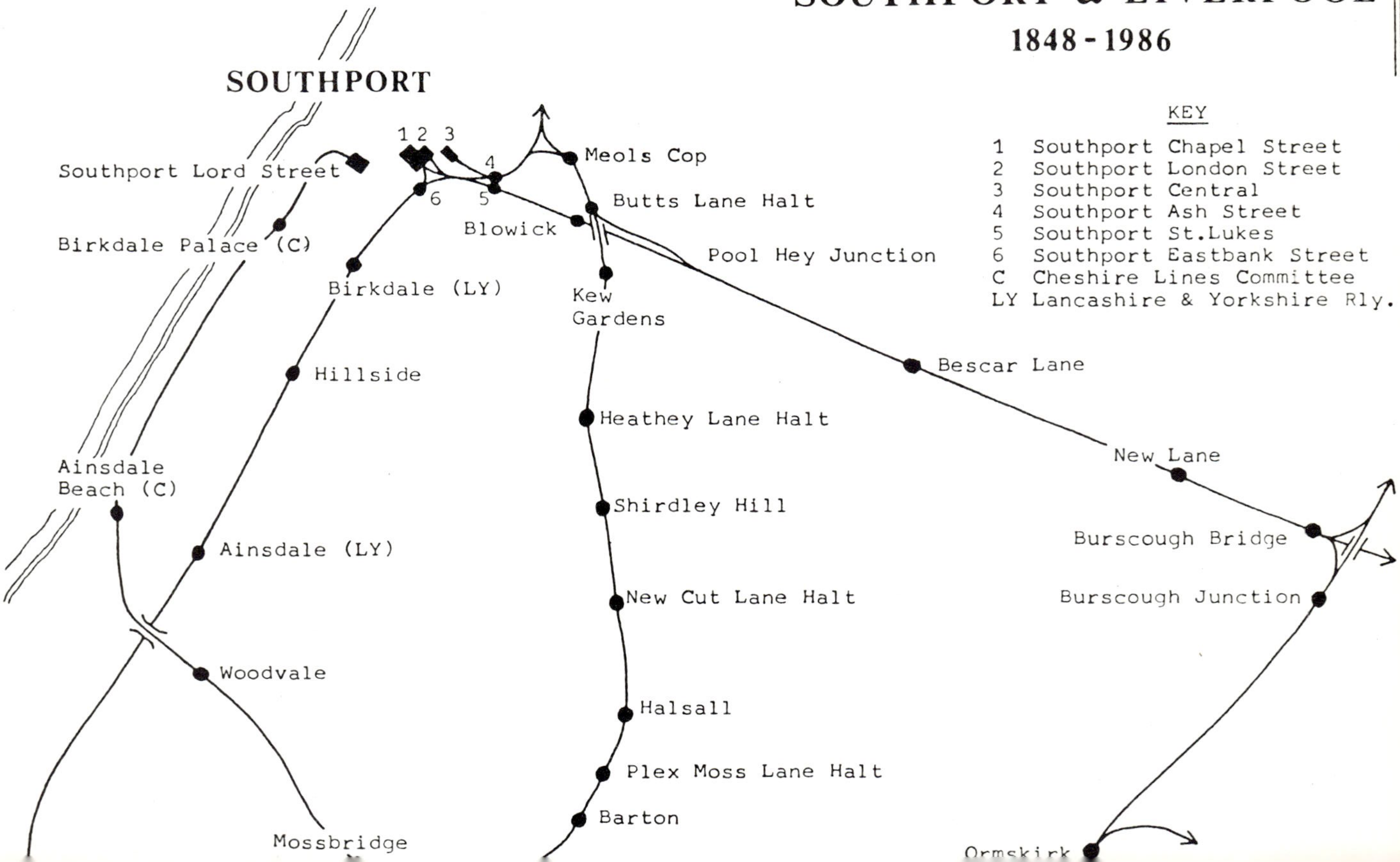

KEY

1 Southport Chapel Street
2 Southport London Street
3 Southport Central
4 Southport Ash Street
5 Southport St.Lukes
6 Southport Eastbank Street
C Cheshire Lines Committee
LY Lancashire & Yorkshire Rly.

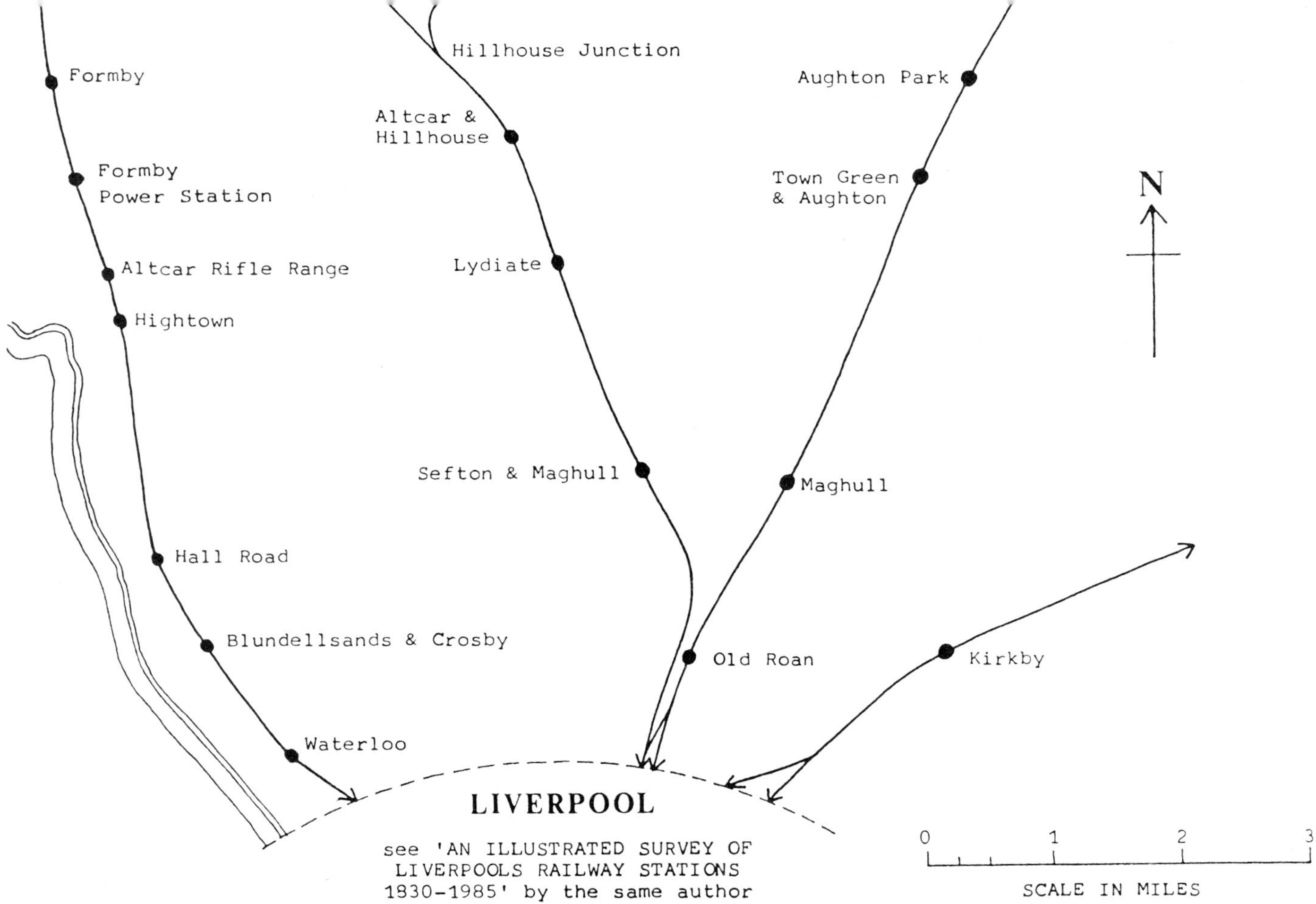
Hillhouse Junction
Formby
Aughton Park
Altcar &
Hillhouse
Formby
Power Station
Town Green
& Aughton
N
Altcar Rifle Range
Lydiate
Hightown
Sefton & Maghull
Maghull
Hall Road
Blundellsands & Crosby
Old Roan
Kirkby
Waterloo
LIVERPOOL
see 'AN ILLUSTRATED SURVEY OF
LIVERPOOLS RAILWAY STATIONS
1830-1985' by the same author
0
1
2
3
SCALE IN MILES

Lydiate (Cheshire Lines Committee) *Line opened* 1-9-1884 *Station opened* 1-9-1884 *Station closed* 1-1-1917 *Station re-opened* 1-4-1919 *Station closed* 7-1-1952 *Line closed* 7-7-1952 The tracks were lifted after June 1960 but only vague traces of the level crossing remain today. The site was North of Carr Lane (continuation of Station Road which is off the A567).

Lydiate station looking North in 1949 (Photo D. Thompson)

Maghull (Lancashire & Yorkshire Railway) *Line opened* 2-4-1849 *Station opened* 2-4-1849 Still open. Electrification reached Maghull in 1909.

Maghull in pre-electrification days looking towards Town Green (Photo P. Woolley collection)

A superb staff pose at Maghull station circa 1910 showing an early electric unit at close quarters (Photo Heyday Publishing Company)

This early view of Maghull shows the station building visible on the left (Photo J. Ryan collection)

Meols Cop (Lancashire & Yorkshire Railway) *Line opened* 1-9-1887 *Station opened* 1-11-1887 Still open. The line was electrified in 1909 but the third rail was unused after 1964 with the closure of the Southport – Crossens service. Once part of the ex-Liverpool, Southport & Preston Junction Railways route to Barton, today the station hosts diverted Southport – Wigan services owing to the closure of the Pool Hey Junc. – St. Lukes section. See Barton notes also.

Meols Cop from the roadbridge in 1965 (Photo D. Thompson)

Meols Cop in 1949 looking towards Southport with a Crossens train visible on the right. Because of the positioning of Meols in relation to Southport and Crossens, trains engaged on this service had to reverse from the station to continue their journey (Photo D. Thompson)

Mossbridge (Cheshire Lines Committee) *Line opened* 1-9-1884 *Station opened* 5-4-1886 as 'Barton & Halsall' being renamed 'Mossbridge' on 1-8-1894 *Station closed* 1-1-1917 *Line closed* 7-7-1952 and the tracks lifted shortly after. It was sited on the South side of Moss Lane which today forms part of the Haskayne – Formby road. The road overbridge still stands but all traces of the station have gone.

Mossbridge from the road photographed after closure (Photo J. Gilmour collection)

Mossbridge in 1937 showing the remains of the Liverpool bound platform (Photo H. B. Priestley)

New Cut Lane Halt (Lancashire & Yorkshire Railway) *Line opened* 1-9-1887 *Station opened* 7-1906 *Station closed* 26-9-1938 *Line closed* 21-1-1952 and the tracks were lifted shortly after. As the below plan shows there were no platforms and the station area stood South of the roadbridge. All traces of the 'Halt' have gone but the bridge still stands (see also Plex Moss Lane view and Barton notes).

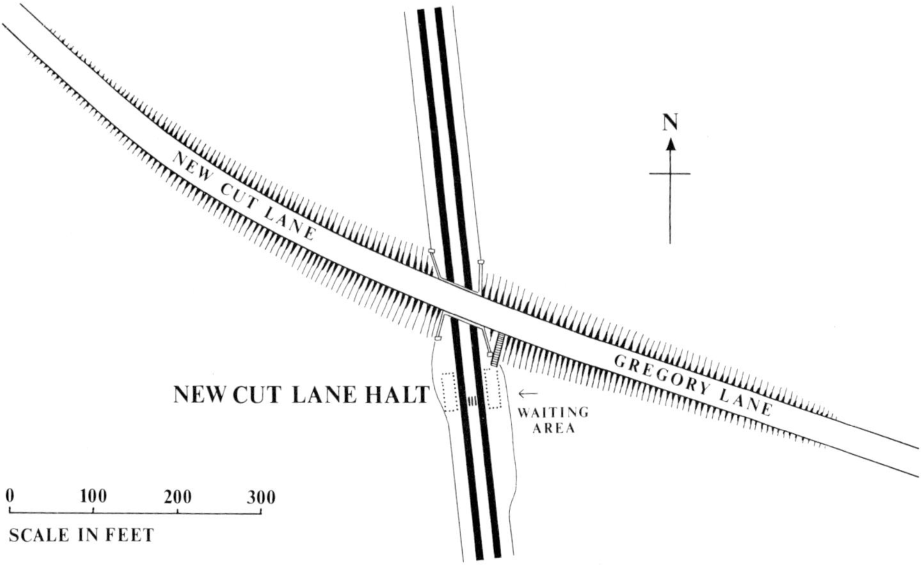

New Lane (Lancashire & Yorkshire Railway) *Line opened* 9-4-1855 *Station opened* 9-4-1855 Still open.

New Lane in 1955 looking East (Photo H. B. Priestley)

New Lanes Southport bound platform in 1957 (Photo H. C. Casserley)

Old Roan (Lancashire & Yorkshire Railway) *Line opened* 2-4-1849 *Station opened* 5-1907 *Station closed* 1-10-1909 *Station re-opened* 1933 Still open. The original station was a halt without platforms being served by a railmotor introduced for economy reasons. Electrification in 1909 brought its closure and it was not until 1933 that a station proper was deemed necessary for the areas growing population.

Old Roan in 1965 looking towards Maghull (Photo D. Thompson)

Ormskirk (Lancashire & Yorkshire Railway) *Line opened* 2-4-1849 *Station opened* 2-4-1849 Still open. Electrification came to the area in 1913. The stations decline began in 1956 with the closure of the Rainford branch but subsequent losses such as the Southport and through Preston – Liverpool services have reduced the station to a skeleton of its former self.

Ormskirk as viewed from the Derby Street bridge circa 1910 (Photo B. Mathews collection)

Ormskirk looking South in early days (Photo Heyday Publishing Company)

The Northern end of Ormskirk in 1955 showing the Rainford branch train on the right with loco No. 41283 (Photo H. B. Priestley)

Plex Moss Lane Halt (Lancashire & Yorkshire Railway) *Line opened* 1-9-1887 *Station opened* 7-1906 *Station closed* 26-9-1938 *Line closed* 21-1-1952 and the tracks lifted shortly afterwards. As the photograph shows there were no platforms and steps were lowered from the coach for passengers access. The station stood on the South side of Plex Moss Lane and steps which led from the road are still visible (see Barton notes also).

The 'Altcar Bob' at Plex Moss Lane Halt circa 1907 (Photo J. Gilmour collection)

Sefton & Maghull (Cheshire Lines Committee) *Line opened* 1-9-1884 *Station opened* 1-9-1884 as 'Sefton' being renamed 'Sefton & Maghull' in 1886 *Station closed* 1-1-1917 *Station re-opened* 1-4-1919 *Station closed* 7-1-1952 *Line closed* 7-7-1952 The tracks were lifted after June 1960 and the site has since been reclaimed for business use. The station stood on the South side of Sefton Lane by the road overbridge.

Sefton & Maghull between the wars looking towards Liverpool (Photo D. Thompson)

Shirdley Hill (Lancashire & Yorkshire Railway) *Line opened* 1-9-1887 *Station opened* 1-11-1887 *Station closed* 26-9-1938 *Line closed* 21-1-1952. Tracks from Butts Lane to this point remained until 1964 for the storage of excursion stock. This ex-Liverpool, Southport & Preston Junction Railway station was situated on Renacre Lane (which is off the A567) but nothing remains now apart from a plaque which marks the site. This was the only station on the Barton branch to have a level crossing instead of a roadbridge (see Barton notes also).

Looking South to the remains of Shirdley Hill in 1949 (Photo D. Thompson)

A conventional passenger train headed by LMS No. 10655 pauses at Shirdley Hill circa 1930 (Photo J. Gilmour collection)

Southport Ash Street (Lancashire & Yorkshire Railway) *Line opened* 10-6-1878 *Station opened* 10-6-1878 as 'Southport Windsor Road' being renamed 'Southport Ash Street' later. Upon opening, this was the temporary wooden terminus of the West Lancashire Railway for trains from Preston. A more appropriate terminus at Southport Central was opened on 5-9-1882 when Ash Street became a ticket platform. *Station closed* 6-1902 when St. Lukes was expanded to serve the Preston branch also (see Southport St. Lukes notes also).

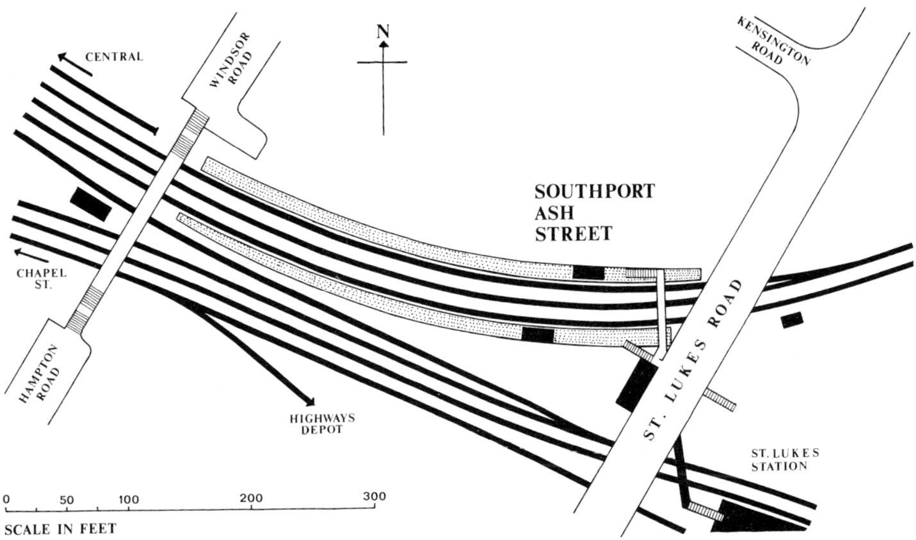

Southport Central (Lancashire & Yorkshire Railway) *Line opened* 5-9-1882 *Station opened* 5-9-1882 *Station closed* 1-5-1901 Upon closure the site became a goods depot and trains from Preston and the Barton branch were re-routed to Chapel Street. Central was the terminus for West Lancashire Railway trains from Preston and was sited at the junction Derby Road and Kensington Road. It survived virtually intact until the 1970's when the land was reclaimed. Today the site is occupied by a commercial enterprise.

The exterior of Southport Central in 1949 (Photo LGRP courtesy of David & Charles Ltd.)

Southport Chapel Street (Lancashire & Yorkshire Railway) *Line opened* 22-8-1851 *Station opened* 22-8-1851 Still open. Originally built as an improved and more centrally located terminus for the single track Liverpool, Crosby & Southport Railway, it replaced the first station at Eastbank Street. This line was doubled in 1852. In 1857 Chapel Street became the terminus for all trains from the Wigan direction, and by 1901, LYR control over its smaller competitors resulted in the services from Preston and Barton terminating here also. Electrification came in 1904 and 1914 saw the expansion of the station to its fullest extent enveloping the land hitherto occupied by the East Lancashire Railway station in London Street (see Southport London Street also). Having been deemed unsafe, the grandiose LYR buildings were replaced by the present day frontage in 1970/1.

LMS railmotor No. 10601 departs Chapel Street in 1928 with a stopping train for Ormskirk. The Barton railmotor (or 'Altcar Bob' as it was affectionately called) also terminated here between 1901 and 1938 (Photo F. Dean)

The exterior of Chapel Street around the turn of the century (Photo C. Loker collection)

Southport Chapel Street from the Victoria footbridge in 1963 showing a number of return special excursions awaiting departure (Photo F. Dean)

Southport Eastbank Street (Lancashire & Yorkshire Railway) *Line opened* 21-7-1848 *Station opened* 21-7-1848 *Station closed* 22-8-1951 Line still open. This was Southports very first railway station being the Northern terminus of the Liverpool, Crosby & Southport Railway. Built of wood and incorporating a roof and waiting rooms, it closed when the line was extended to Chapel Street.

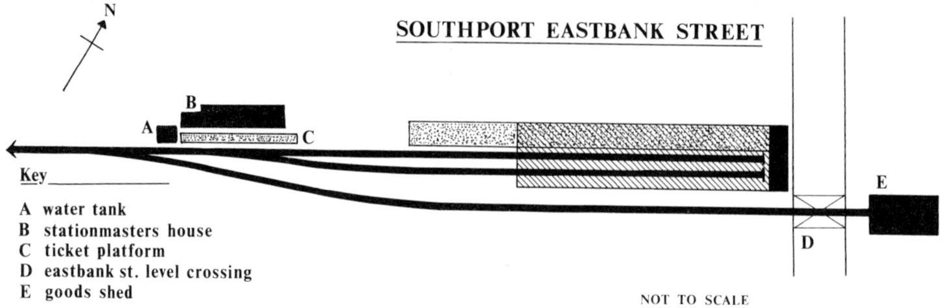

Southport London Street (Lancashire & Yorkshire Railway) *Line opened* 9-4-1855 *Station opened* 9-4-1855 *Station closed* 1-4-1857 Line still open. This was the East Lancashire Railways terminus for the LYR/ELR jointly built Wigan line (the LYR's terminus being Chapel Street which stood alongside). The two companies amalgamated in 1859. Upon closure the services transferred to Chapel Street and the ELR building became a goods depot. The 1914 expansion of Chapel Street swallowed the London Street site but it's name was preserved with the provision of excursion platforms on more or less the same land. It has to be said, however, that the 'London Street Excursion Platforms' were merely Nos. 12 & 13 of Chapel Street.

London Street excursion platforms as viewed from the Victoria footbridge in 1965 (Photo D. Thompson)

Southport Lord Street (Cheshire Lines Committee) *Line opened* 1-9-1884 *Station opened* 1-9-1884 *Station closed* 1-1-1917 *Station re-opened* 1-4-1919 *Station closed* 7-1-1952 *Line closed* 7-7-1952 The tracks were lifted shortly after and the curved approach line has since been reclaimed as a road. The station building still stands at the junction of Lord Street West and Duke Street and is in use as a bus terminus.

The approaches to Lord Street in 1951 with a Liverpool train headed by No. 40397 about to leave (Photo F. Dean)

The interior and buffer-stops at Lord Street in 1949 (Photo D. Thompson)

A view from Lord Street's concourse in 1951 (Photo F. Dean)

Lord Street exterior around the turn of the century (Photo Heyday Publishing Company)

Southport St. Lukes (Lancashire & Yorkshire Railway) *Line opened* 9-4-1855 *Station opened* 2-7-1883 as 'Barton Street' It was renamed 'Southport (St. Lukes Road)' circa 10-1883, 'Southport (St. Lukes)' on 1-3-1902 and finally 'St. Lukes' about 5-1914. The adjacent Southport Ash Street was absorbed by St. Lukes in June 1902 when the latter began serving the Preston branch also. Electrification was introduced in 1904 (Preston line only). *Station closed* (original portion) 14-6-1965 (Preston portion) 8-9-1968. Few traces remain today (see Southport Ash Street notes also).

St. Lukes (Wigan platforms) in 1962 with loco No. 42299 arriving at the head of an ex-Southport train (Photo H. B. Priestley)

St. Lukes (Preston platforms) in 1965 looking in the direction of Meols Cop (Photo D. Thompson)

Town Green & Aughton (Lancashire & Yorkshire Railway) *Line opened* 2-4-1849 *Station opened* 2-4-1849 as 'Town Green' being renamed 'Town Green & Aughton' on 1-6-1889. It reverted to 'Town Green' on 5-5-1975. Electrification came to the area in 1911. Still open.

Town Green & Aughton circa 1905 looking North (Photo Heyday Publishing Company)

Waterloo (Lancashire & Yorkshire Railway) *Line opened* 21-7-1848 *Station opened* 21-7-1848 Still open. This venue was once the single track terminus of the Liverpool, Crosby & Southport Railway prior to the building of the Liverpool extension (opened 1-10-1850). Hitherto, omnibuses plied between Waterloo and Liverpool to complete the connection. The line was doubled in 1852 and the station rebuilt in 1880/1. A road bridge was also provided at this time to replace the level-crossing. The line was electrified in 1904. Waterloo was the scene of a tragic accident on 15-7-1903 when a Southport bound express became derailed and mounted the platform killing eight and injuring a further 20 people.

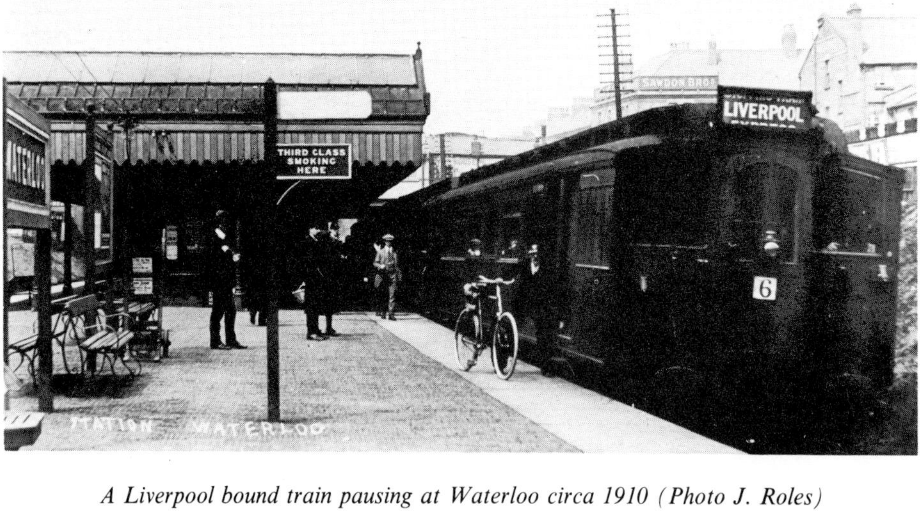

A Liverpool bound train pausing at Waterloo circa 1910 (Photo J. Roles)

Waterloo's Southport bound platform in early days (Photo Heyday Publishing Company)

Waterloo looking North in 1965 (Photo D. Thompson)

Woodvale (Cheshire Lines Committee) *Line opened* 1-9-1884 *Station opened* 1-9-1884 as 'Woodville & Ainsdale' being renamed 'Woodvale' on 1-5-1898. *Station closed* 1-1-1917 *Station re-opened* 1-4-1919 *Station closed* 7-1-1952 *Line closed* 7-7-1952 The tracks were lifted shortly after and the trackbed from the site towards Southport has since been utilised as the coastal road. The station was located on Coastal Road at its Junction with the A565 but nothing remains now.

Woodvale circa 1910 looking towards Southport (Photo Heyday Publishing Company)

Another early view of Woodvale but looking in the opposite direction (Photo J. Gilmour collection)